IS FOR HOPE

CLIMATE CHANGE FROM A to Z

IS FOR HOPE

ELIZABETH KOLBERT

ILLUSTRATIONS BY WESLEY ALLSBROOK

ONEWORLD

ARRHENIUS
8

BLAH, BLAH, BLAH
14

CAPITALISM
22

DESPAIR
28

ELECTRIFY EVERYTHING
32

FLIGHT
38

GREEN CONCRETE
44

HOPE
50

INFLATION REDUCTION ACT
56

CONTENTS

JOBS, JOBS, JOBS
62

KILOWATT
66

LEAPFROGGING
70

MATH
76

NARRATIVES
80

OBJECTIONS
86

POWER
92

QUAGMIRE
96

REPUBLICANS
102

SHORTFALL
106

TEMPERATURES
112

UNCERTAINTY
120

VAST
124

WEATHER
130

XENOPHOBIA
136

YOU
142

ZERO
148

ACKNOWLEDGMENTS 156

ABOUT THE CONTRIBUTORS 157

FURTHER READING 158

Climate change resists narrative—and yet some account of what's happening is needed.

Millions of lives are at stake. And upward of a million species.

And there are decisions to be made, even if it's unclear who, exactly, will make them.

ARRHENIUS

Svante Arrhenius was, by nature, an optimist. He believed that science should—and could—be accessible to all. In 1891, he got his first teaching job, at an experimental university in Stockholm called the Högskola. That same year, he founded the Stockholm Physics Society, which met every other Saturday evening. For a fee of one Swedish crown, anyone could join. Among the society's earliest members was a Högskola student named Sofia Rudbeck, who was described by a contemporary as "an excellent chemist" and "a ravishing beauty." Arrhenius began writing her poetry, and soon the two wed.

Physics Society meetings consisted of lectures on the latest scientific developments, many delivered by Arrhenius himself. These were followed by discussions that often lasted well into the night. The topics ranged widely, from aeronautics to volcanology. The society devoted several sessions to considering the instruments that would be needed by Salomon August Andrée, another early member of the group, who had decided to try to reach the North Pole via balloon. (Whatever the quality of his instruments, Andrée's voyage would result in his own death and the death of his two companions.)

A question that particularly interested the Physics Society was the origin of the ice ages. All over Sweden lay signs of the glaciers that had, for vast stretches of time, buried the country: rocks with parallel scrapings; strange, sinuous piles of gravel; huge boulders that had been transported far from their source. But what had caused the great ice sheets to descend, carrying all before them? And then what had caused them to retreat, allowing the rivers to flow once again and the forests to return? In 1893, the society debated various theories that had been proposed, including one linking the ice ages to slight variations in the Earth's orbit. The following year, Arrhenius came up with a different—and, he thought, better—idea: carbon dioxide.

A

Carbon dioxide, he knew, had curious heat-trapping properties. In the atmosphere, it allowed visible light to pass through, but it absorbed the longer-wave radiation that the Earth was constantly emitting to space. What if, Arrhenius speculated, the amount of CO_2 in the air had varied? Could that explain the glaciers' ebb and flow?

The math involved in testing this theory went far beyond what was possible at the time. Arrhenius didn't have a calculator, let alone a computer. He lacked crucial information about which wavelengths, exactly, CO_2 absorbs. The climate system, meanwhile, is immensely complicated, with feedback loops nestled within feedback loops.

Arrhenius, who would later win a Nobel Prize for an unrelated discovery, plunged ahead anyway. On Christmas Eve, 1894, he began constructing a climate model—the world's first. He assembled temperature data from around the globe and made ingenious use of a set of measurements that had been taken a decade earlier by an American astronomer, Samuel Pierpont Langley. (Langley had invented a device—a bolometer—for gauging infrared radiation and had used it to determine the temperature of the moon.) Arrhenius performed thousands of computations—perhaps tens of thousands—and often labored over this task for fourteen hours a day. He was still calculating away as his marriage fell apart. In September of 1895, Rudbeck moved out. In November, without having seen Arrhenius again, she gave birth to their son. The following month, Arrhenius finished his work.

"I should certainly not have undertaken these tedious calculations if an extraordinary interest had not been connected with them," he wrote.

Arrhenius believed that he had unravelled the mystery of the ice ages, a riddle that had "hitherto proved most difficult to interpret." He was at least partly right: ice ages are the product of a complex interplay of forces, including wobbles in the Earth's orbit *and* changes in atmospheric CO_2.

A

His model turned out to have another use as well. All across Europe and North America, coal was being shovelled into furnaces that were bellowing out carbon dioxide. By thickening the atmospheric blanket that warmed the Earth, humans must, Arrhenius reasoned, be altering the climate. He calculated that if the amount of carbon dioxide in the air were to double, then global temperatures would rise between three and four degrees Celsius. A few quadrillion computations later, vastly more advanced climate models predict that doubling CO_2 will push temperatures up between 2.5 and four degrees Celsius, meaning that Arrhenius's pen-and-paper estimate was, to an uncanny degree, on target.

Arrhenius thought that the future he had conjured would be delightful. "Our descendants," he predicted, would live happier lives "under a warmer sky." The prospect was, in any event, distant; doubling atmospheric CO_2 would, he reckoned, take humanity three thousand years.

It's easy now to poke fun at Arrhenius for his sunniness. The doubling threshold could be reached within decades, and the results of this are apt to be disastrous. But who among us is really any different? Here we all are, watching things fall apart. And yet, deep down, we don't believe it.

BLAH, BLAH, BLAH

On September 28, 2021, at the Youth4Climate conference, held in Milan, Greta Thunberg took the stage. Sitting near her was the city's mayor, Giuseppe Sala, wearing a mask. Thunberg, who is five feet tall, could barely be seen over the lectern. She had removed her mask and was smiling.

"Climate change is not only a threat, it is above all an opportunity to create a healthier, greener, and cleaner planet which will benefit all of us," she began. "We must seize this opportunity—we can achieve a win-win in both ecological conservation and high-quality development. . . . We need to walk the talk; if we do this together, we can do this.

"When I say 'climate change,' what do you think of?" she went on. "I think of jobs—green jobs." This received a round of applause.

"We must find a smooth transition towards a low-carbon economy," Thunberg said. "There is no Planet B. There is no Planet Blah—blah, blah, blah; blah, blah, blah." Her listeners, including Sala, started to realize that they'd been had. The applause died down.

"Build Back Better—blah, blah, blah," Thunberg continued.

"Green economy—blah, blah, blah.

"Net zero by 2050—blah, blah, blah.

"Net zero—blah, blah, blah.

"Climate neutral—blah, blah, blah.

"This is all we hear from our so-called leaders: words—words that sound great, but so far have led to no action," Thunberg said. "Of course we need constructive dialogue, but they've now had thirty years of blah, blah, blah, and where has that led us?"

B

Despite decades of promises, global CO_2 emissions have continued to rise.

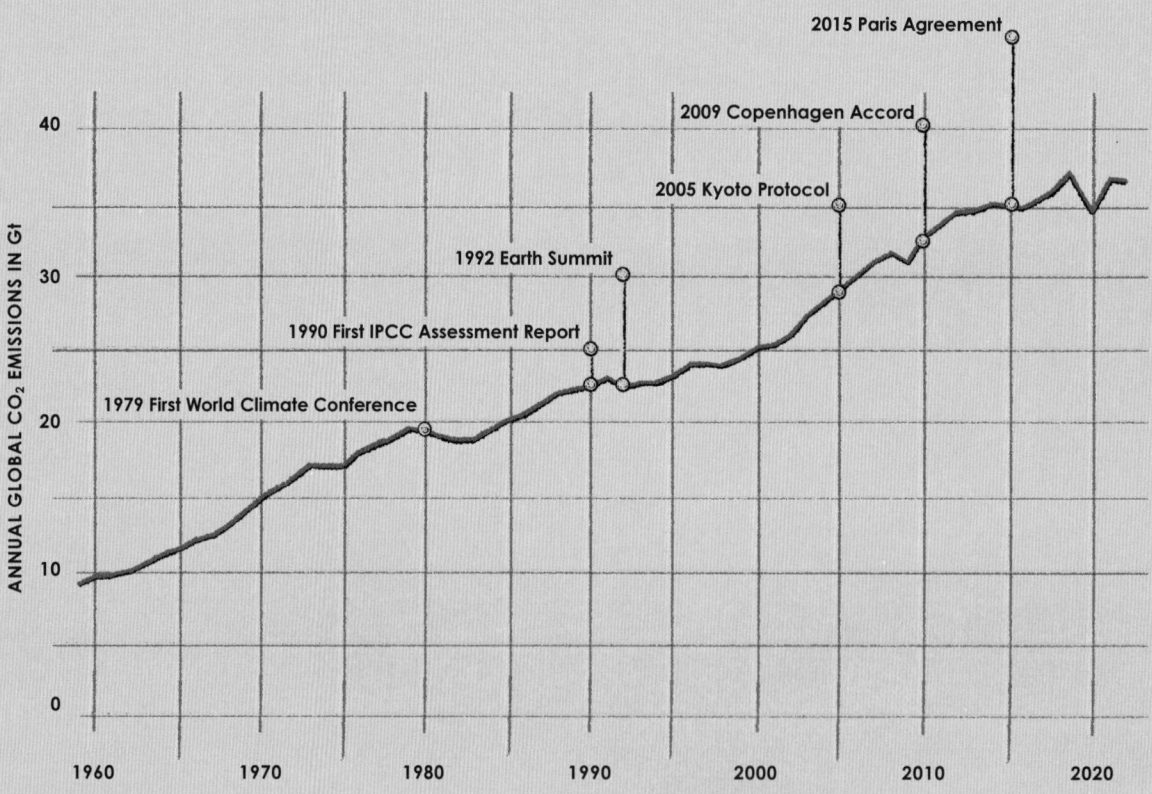

B

It was in 1992 that the world's "so-called leaders" gathered in Rio de Janeiro for the so-called Earth Summit. Everyone agreed that radical change was needed. To avert disaster, global CO_2 emissions, which were then running at around twenty-two billion metric tons a year, would have to be reduced, eventually almost to zero. How this would happen, no one really knew. Still, the goal of preventing "dangerous" warming was enshrined in the United Nations Framework Convention on Climate Change, which President George H. W. Bush cheerfully signed.

"Some find the challenges ahead overwhelming," Bush said. "I believe that their pessimism is unfounded."

A follow-up conference of the parties, or COP, took place in Kyoto five years after the Earth Summit, in 1997. By that point, annual global emissions had risen to twenty-four billion tons. After much back-and-forth, it was agreed that something had to be done. The Kyoto Protocol, an addendum to the Framework Convention, laid out specific emissions-reduction targets for countries to meet.

"I am both determined and optimistic that we can succeed," Vice President Al Gore told the diplomats gathered in Japan.

After Kyoto, global emissions kept on rising, only faster. By 2009, they'd climbed to thirty-two billion tons a year. That fall, President Barack Obama flew to Copenhagen for yet another conference of the parties—COP15. "I believe that we can act boldly, and decisively, in the face of this common threat," he declared.

By 2015, emissions had increased to thirty-five billion tons a year. At that year's COP—No. 21—held in Paris, it was decided that, at last, really and truly, it was time to get serious. "The decisions you make here will reverberate down through the ages," the United Nations secretary-general, Ban Ki-moon, told the delegates. Nevertheless, emissions continued to rise. In the past thirty years, humans have added as much CO_2 to the atmosphere as they did in the previous thirty thousand.

At some point during all the "blah, blah, blah"-ing—it's hard to say when, exactly—climate change ceased to be a prospective problem and became a clear and present one. Since Rio, the Arctic ice cap has shrunk by two-fifths. Greenland has shed some four trillion metric tons of ice, and mountain glaciers have lost six trillion tons. Heat waves are now hotter, droughts deeper, and storms more intense. In some parts of the world, the wildfire season never ends.

One conclusion to draw from this pattern is that the world isn't going to avoid "dangerous" warming. Global leaders will continue to gather at COPs and to speak loftily about "net zero" and "a low-carbon economy." But nothing will change, and, as a result, everything will change. There will be large-scale crop failures. The Greenland ice sheet will start to collapse—it may already be collapsing—and, owing to sea-level rise in some places and desertification in others, large swaths of the globe will become uninhabitable.

This conclusion is not, however, the one that Thunberg chose to draw when she spoke at the Youth4Climate conference. "Right now we are still very much speeding in the wrong direction," she told the crowd in Milan. "But, of course, we can still turn this around—it is entirely possible.

"The leaders like to say, 'We can do this,'" she went on. "They obviously don't mean it, but *we* do—we can do this. I'm absolutely convinced that we can."

Or, as Thunberg might put it, Blah, blah, blah.

B

CAPITALISM

What's the matter here? Why has so little progress been made on climate change, even as the dangers have become ever more apparent?

According to one school of thought, the problem has to do with incentives. There's a great deal of money to be made selling fossil fuels—in 2022, the world's five largest oil-and-gas companies announced profits of almost two hundred billion dollars—and still more money to be made by burning fossil fuels to make stuff to sell, from sunglasses to steel girders. Meanwhile, the costs of climate change can be fobbed off on someone else. To use the technical term, they are a "negative externality." In the words of the Stern Review, a report commissioned by the British government in 2005, climate change "is the greatest and widest-ranging market failure ever seen."

By this account, the obvious solution is to realign the incentives—to internalize the externalities. If the cost of the damage caused by a ton of CO_2 was borne by the business (or individual) responsible for emitting that ton, then the business (or individual) would be motivated to cut back. "A carbon tax offers the most cost-effective lever to reduce carbon emissions at the scale and speed that is necessary," a 2019 statement signed by thirty-five hundred economists, including twenty-eight Nobel Prize winners, declared. Such a tax would move "the invisible hand of the marketplace to steer economic actors towards a low-carbon future."

C

According to a second school of thought, the trouble runs a whole lot deeper. Our political system is dominated by corporate money in general and fossil-fuel money in particular. (The oil-and-gas industry reports spending more than a hundred and twenty million dollars a year lobbying Washington, and probably it spends a great deal more via front groups.) It's therefore naive to imagine that policies that cut into fossil-fuel profits will be enacted. And even if they were, they wouldn't solve the essential problem, which is that the "invisible hand" always grasps for more. If it's not more oil, it will be more lithium to build batteries, and if it's not more lithium, it will be more cobalt, mined from the bottom of the sea.

"When it comes to global warming, we know that the real problem is not just fossil fuels—it is the logic of endless growth that is built into our economic system," Jason Hickel, an economic anthropologist at the Autonomous University of Barcelona, has written. Climate change can't be dealt with using the tools of capitalism, because it is a product of capitalism. It can be dealt with only by throwing off capitalism in favor of something else—a system aimed not at growth but at "degrowth."

"The difficult truth is that, to prevent climate and ecological catastrophe, we need to level down" is how the British environmental writer George Monbiot has put it.

A third line of thought—perhaps too bleak and unpopular to be called a school—is that, if big change is hard, bigger change is even harder. How are we going to build a whole new economic system if we can't even enact a carbon tax?

DESPAIR

Despair is unproductive.

D

It is also a sin.

ELECTRIFY EVERYTHING

Let's try again, this time with feeling.

BIWF2 is a wind turbine that sticks up out of the Atlantic Ocean, about fifteen miles off the coast of Rhode Island. It's six hundred feet tall, which is higher than the Washington Monument, and its blades are more than two hundred feet long. I'm on a boat designed to transport crews to offshore wind farms. The captain maneuvers right up to the metal stanchions that hold the turbine in place, so the blades are rotating directly overhead. They make a fantastic whooshing sound that builds and fades, builds and fades. The effect is at once thrilling and terrifying, as if some gigantic bird were trying to land on the deck. "Ah," everyone on board exclaims as another blade descends.

BIWF2 has one neighbor half a nautical mile to the north and three more neighbors to the south. Together the turbines make up Block Island Wind Farm, America's first offshore wind operation. A dozen more wind projects are currently planned off the East Coast, from Massachusetts to North Carolina. The turbines that will be erected in these projects will make BIWF2 look puny.

Staring up at the blades, I am looking into the future—or at least *a* future— and it's inspiring. BIWF2 is a symbol of what can be accomplished when people put their minds to it.

E

In 1992, the year of the Earth Summit, the world had exactly one offshore wind farm, called Vindeby. Situated off the Danish island of Lolland, it consisted of eleven turbines, which, collectively, produced less power than BIWF2 does today. Now there are scores of offshore farms, most of them in European and Chinese waters. The largest, known as Hornsea 2, is in the North Sea, off the English coast; it comprises a hundred and sixty-five turbines, each so massive that a single sweep of its blades can power a household for a day. Block Island Wind Farm and Hornsea 2 are owned by the same company, which used to be known as Danish Oil and Natural Gas, or DONG, but recently—and for obvious reasons—changed its name, to Ørsted. (It also owned Vindeby, which was decommissioned in 2017.) As more turbines have gone up, costs have plunged; just in the past decade, the price of offshore wind energy has declined by half.

Onshore wind has grown even faster, and its cost has also plummeted. In many parts of the world, it's now cheaper to put up turbines than it is to operate an existing power plant that burns natural gas. In places with a lot of wind, such as Denmark, Ireland, and western Oklahoma, there's sometimes so much power pouring into the grid that producers have to pay to get rid of it.

The price of solar power, meanwhile, has declined even more spectacularly. Since 2010, it's dropped by more than eighty per cent. According to the International Energy Agency, solar power now offers "some of the lowest-cost electricity ever seen."

The rapidly falling price of renewables makes it possible to imagine a not too distant future in which the United States, indeed the world, generates all its electricity emissions-free. Already there are brief periods—on the order of minutes—when California can produce enough electricity from renewables to meet its demand. In Denmark, this happens for entire windy days.

And, once it's possible to imagine a carbon-free grid, all sorts of other opportunities open up. Substitute electric motors for internal-combustion engines and cars, too, can run emissions-free. The same goes for trucks and buses, ferries and forklifts. Plug them in! Tear out boilers and replace them with heat pumps! Swap gas ranges for induction stoves! Electrify as much as possible. Ideally, electrify everything.

E

The cost of renewable energy has been falling dramatically.

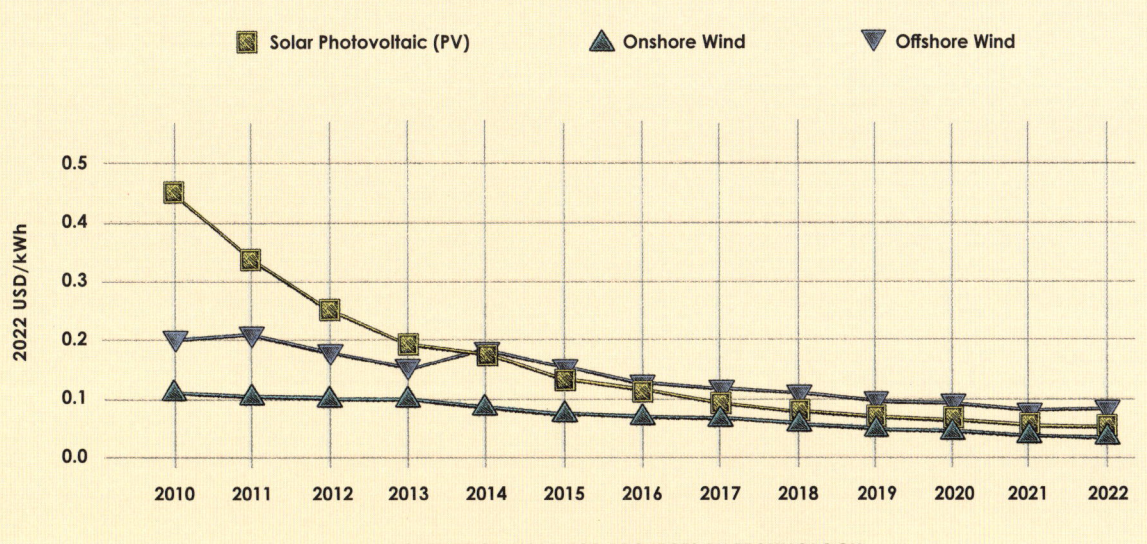

Source: IRENA

FLIGHT

The Alia, a sleek, all-white aircraft with angular wings, looks like the offspring of an F-16 and a seabird. It has four propellers mounted on poles that bisect the wings horizontally; these rotate ceiling-fan style. I'm sitting in the cockpit of a mockup of the plane that is used to train pilots. My flight instructor, Vince Moeykens, directs me to pull, very gently, on a lever attached to the floor. The four propellers start to spin, the floor vibrates, and the surprisingly convincing simulacrum of the ground that's being projected onto a screen in front of us begins to fall away.

"We're really starting to climb," Moeykens says, encouragingly. I lower the lever; the propellers on the top of the plane stop spinning, and one in the back takes over. We are now ostensibly moving forward, toward a large lake, which, since we're in Burlington, Vermont, I take to be Lake Champlain. Moeykens assures me that I am doing a great job, but soon the ground appears to be whizzing back toward us, and I tug on another lever to lift the plane's nose. I steer us out toward the simulated lake, and we glide peacefully over the simulated water. Just as I think I am starting to get the hang of things, it's time to turn around. I bank too sharply and, remarkably convincingly, the plane seems to pitch. As I attempt to land, the aircraft lurches, and we have a near-miss with what looks like a parking garage. The queasiness I feel on stumbling out of the cockpit is entirely genuine.

The Alia runs solely on electricity. It's manufactured by a company called Beta, which was founded, in 2017, by a former professional hockey player named Kyle Clark. (Clark studied mechanical engineering as an undergraduate, at Harvard; while there, he wrote a thesis about a plane that could be piloted like a motorcycle.) Beta occupies a U-shaped building overlooking a hangar at the Burlington airport. On the day I visited, several planes were parked in the hangar, including an Alia. The other Alia—at the time, only two existed—was wending its way back to Vermont from Bentonville, Arkansas, where it had been flown, in several hops, as a test. Clark had piloted the plane on some of the return-trip hops.

F

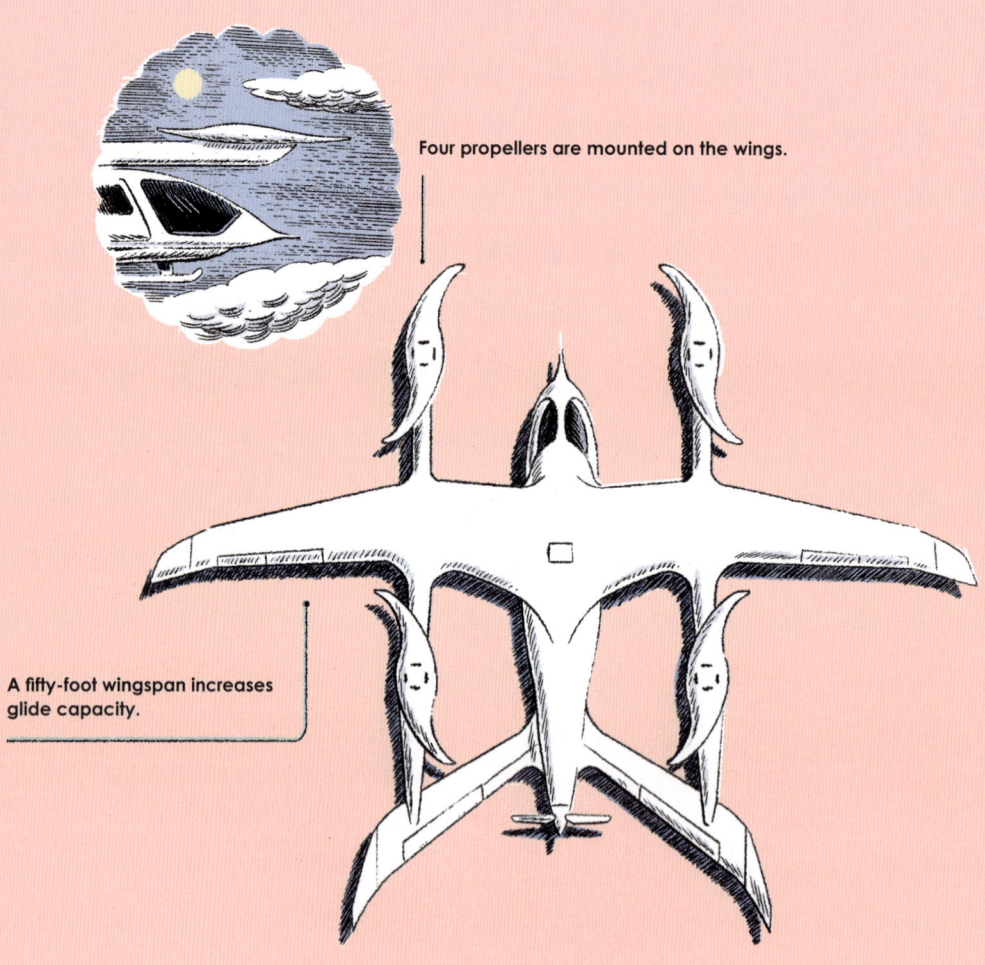

The Beta Alia flies on stored electricity.

Four propellers are mounted on the wings.

A fifty-foot wingspan increases glide capacity.

Source: Beta Technologies

"Each one of those legs was about a hundred and eighty miles," he told me. We were sitting in a conference room one floor up from the flight simulator, overlooking the hangar. The Alia, Clark explained, had been shadowed by a Cessna Caravan carrying a crew collecting data. On each leg, Clark estimated, the Alia had gone through about twenty dollars' worth of stored electricity. To travel the same distance, the Cessna had burned about six hundred dollars' worth of jet fuel.

"The reason I bring that up," Clark continued, "is that, ultimately, what we are paying for in those two cases is energy. And the difference is a product of how much you're wasting."

Compared with internal-combustion engines, electric motors are extremely efficient; an electric car can travel at least twice as far as a conventional sedan on the same amount of energy. The challenge for electrifying flight is that batteries are heavy. (Jet fuel, too, is heavy, but it's a lot more energy dense, plus its weight drops as it gets burned.)

Yet, as the Alia shows, rapid progress is being made. Beta has applied to the Federal Aviation Administration for approval to transport cargo, and UPS has already placed an order for up to a hundred and fifty Alias. There are also electric passenger planes in the works. Eviation, a company headquartered in Washington State, has produced a prototype of a nine-seater, dubbed Alice. Cape Air, based in Hyannis, Massachusetts, has said it intends to buy seventy-five Alices. Wright Electric, headquartered in Albany, New York, is planning to transform a British Aerospace commuter jet, the BAe 146, into an electric hundred-seater. "All short flights can be zero-emissions starting in 2026," Wright's Web site proclaims.

GREEN CONCRETE

"We are doing freeze-and-thaw tests here in this lab," Mehrdad Mahoutian said. He pried the lid off a plastic container of the sort usually used to store leftovers. Inside was a gray block about the size of a juice box. It was sitting in a half inch or so of ice-fringed water.

"This is cement-free concrete," Mahoutian said, indicating the block. "And this is salt water. For eighteen hours, they go into the freezer. And, for six hours, they get melted, basically." It was the day after my journey in the flight simulator, and, following an unfortunate delay at the Canadian border, I had managed to find the headquarters of a company named CarbiCrete, in an industrial area of Montreal. Mahoutian, one of the company's founders, was showing me around the R. & D. facility. Every few minutes, he was interrupted by a very loud rumble. "That's the blocks being made," he shouted over the din.

We passed into a second room, where two test walls of cinder block stood perpendicular to each other. Both were equipped with a shower apparatus made from PVC pipe, which was dripping water. A fan blew the water toward the blocks. Mahoutian explained that one test wall had been constructed with ordinary cinder blocks, the other with a new kind of block fabricated by CarbiCrete. The shower arrangement was gauging how the two walls compared in terms of water penetration. "In a few hours, we'll measure the dampness and do some calculations," he told me.

Concrete represents one of the world's most obdurate carbon problems. Its key ingredient, Portland cement, is made by grinding up limestone, adding clay, and heating the mixture to more than two thousand degrees Fahrenheit. The process demands a lot of energy, which is usually supplied by burning coal. But, more fundamentally, the issue with cement is its chemistry; heating limestone to the point that it transforms into quicklime unavoidably releases CO_2. In 2023, some thirty billion tons of concrete were produced worldwide, almost four tons for every single person on the planet. The associated carbon emissions accounted for roughly eight per cent of the global total—more than aviation and shipping combined. Producing cement-free concrete, or what is sometimes referred to as green concrete, isn't sexy, but it's essential.

In place of cement, CarbiCrete makes use of a waste product—the slag left over from steel production. It pounds the slag into powder and mixes in crushed rock and water. The resulting slurry, which looks a lot like conventional concrete, can then be molded into blocks or tiles.

CarbiCrete bills its product, which for the time being is also known as CarbiCrete, not just as carbon-neutral but as carbon-negative. Mahoutian led me to a row of machines that resembled rice cookers. Each one was attached to a cannister of CO_2. Inside the machines, little blocks of damp CarbiCrete were reacting with carbon dioxide; instead of releasing the gas, the blocks were soaking it up.

"Please touch," Mahoutian instructed. The machines were hot. This, he explained, was because the reaction, rather than requiring heat, generated it.

For now, CarbiCrete buys its CO_2 from a supplier. The plan, though, is eventually to use carbon that's been captured at, say, a power plant or a steel mill.

"What we are doing basically is killing three birds with one stone," Mahoutian told me. "We are not using cement. We are permanently capturing CO_2. And we're reducing the need for landfills." As I was getting ready to leave, Mahoutian asked if I wanted a CarbiCrete tile or cinder block to bring home with me. I thought for a while and then decided to take both.

HOPE

"Hope is the pillar that holds up the world," Pliny the Elder is supposed to have observed. "Hope is the dream of a waking man." Go looking for hopeful climate stories and they turn up everywhere.

Not long ago, I came across one in a defunct wine distributorship, in Somerville, Massachusetts. The cavernous warehouse had been taken over by a company called Form Energy, whose waking dream concerns rust. Rusting usually proceeds in one direction, and the end result is a corroded nail or screw that winds up in the trash. But, as iron oxidizes, it gives up electrons. Therefore, if a current is applied to rust in solution, the process will run in reverse. At Form, the goal is to use this reverse-rusting trick to make a new kind of battery, one so cheap and durable it could power an entire city.

Billy Woodford, Form's chief technology officer, studied material science at M.I.T. "Batteries have cool technical problems," he told me as we descended into the warehouse turned research lab. The huge room was lined with experimental chambers that resembled glass-fronted refrigerators. Each was labelled, according to an inside joke that I never quite got, with the name of a different Oreo variety, like lemon or s'mores or gluten free. (I was surprised to learn that there had been a Lady Gaga Oreo, which consisted of pink wafers and green cream.)

Inside the chambers were collections of some kind of high-tech Tupperware, with wires poking through the lids. The containers, in turn, held plates of iron bathing in liquid. Woodford explained that these were test batteries: "We'll put in different iron—there's different versions, depending on whether it's produced, say, in Texas or Germany—and then different electrolytes."

H

⌐ **Iron-air batteries supply power as they rust.**

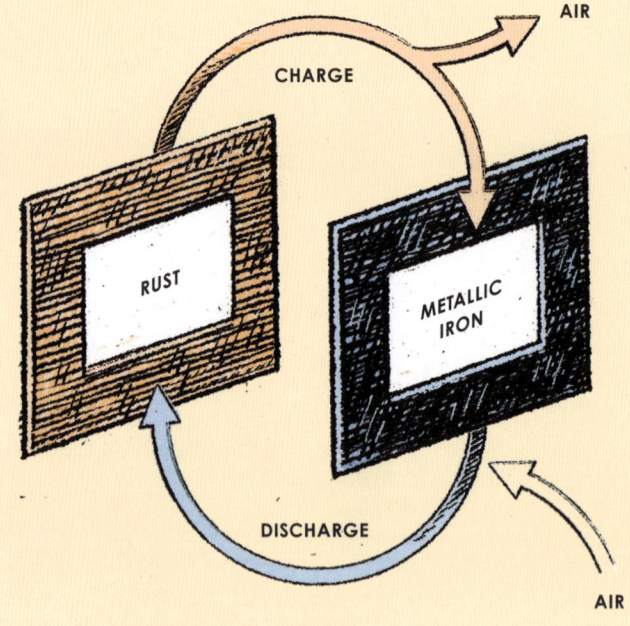

Form's full-scale batteries are going to be packaged in modules of fifty. Each module will be about the size of a washer and dryer placed side by side, and ten of them will be big enough to fill a shipping container. The first thirty shipping containers' worth have been promised to Great River Energy, a Minnesota-based utility that buys a lot of wind power. The idea is that on blustery days the batteries will soak up energy. Then, on calm days, the plates will rust and feed current back into the grid.

Form's C.E.O., Mateo Jaramillo, studied theology and later became a Tesla executive. While at Tesla, he worked on lithium-ion batteries, which are the sort used in most electric vehicles (and in the Alia), and also, in a slightly different form, in laptops and cell phones.

"Lithium-ion is fantastic," Jaramillo told me. "And yet, if that's the only tool you have, you still have a really hard time replacing high-capacity coal and natural-gas plants. To replace those, you need something that's at least an order of magnitude cheaper than lithium-ion." The materials needed for reversible rusting—air, salt water, and iron—are available in practically limitless quantities. "Besides coal, iron is the most-mined mineral on earth," Jaramillo said. "So it scales."

INFLATION REDUCTION ACT

"I can say with confidence that I have never been more optimistic about America." So tweeted President Joe Biden shortly after signing the Inflation Reduction Act, on August 16, 2022. The bill, which authorizes federal spending of more than three hundred and fifty billion dollars on climate initiatives, had, Lazarus-like, sprung from the dead a month earlier, stunning just about everyone, including most members of the U.S. Senate.

Biden came into office vowing sweeping action on climate change. His first day in the White House, he announced that the United States was rejoining the Paris Agreement, from which it had withdrawn under Donald Trump. Biden subsequently pledged to the world that, by 2030, the U.S. would cut its emissions by fifty per cent, compared with 2005.

Biden's pledge wasn't quite as ambitious as it seemed. The U.S.'s emissions peaked around 2005; since then, thanks largely to a shift from coal to natural gas, they have declined by about twenty per cent. Still, Biden's promise was a reach. To have any hope of making good on it, the president needed to get legislation through Congress. To get legislation through Congress, he needed all fifty Senate Democrats. And to get all fifty Democrats, he needed Joe Manchin.

It is estimated that by 2035, U.S. emissions will be thirty-two to fifty-one per cent lower than they were in 2005, owing in large part to the Inflation Reduction Act.

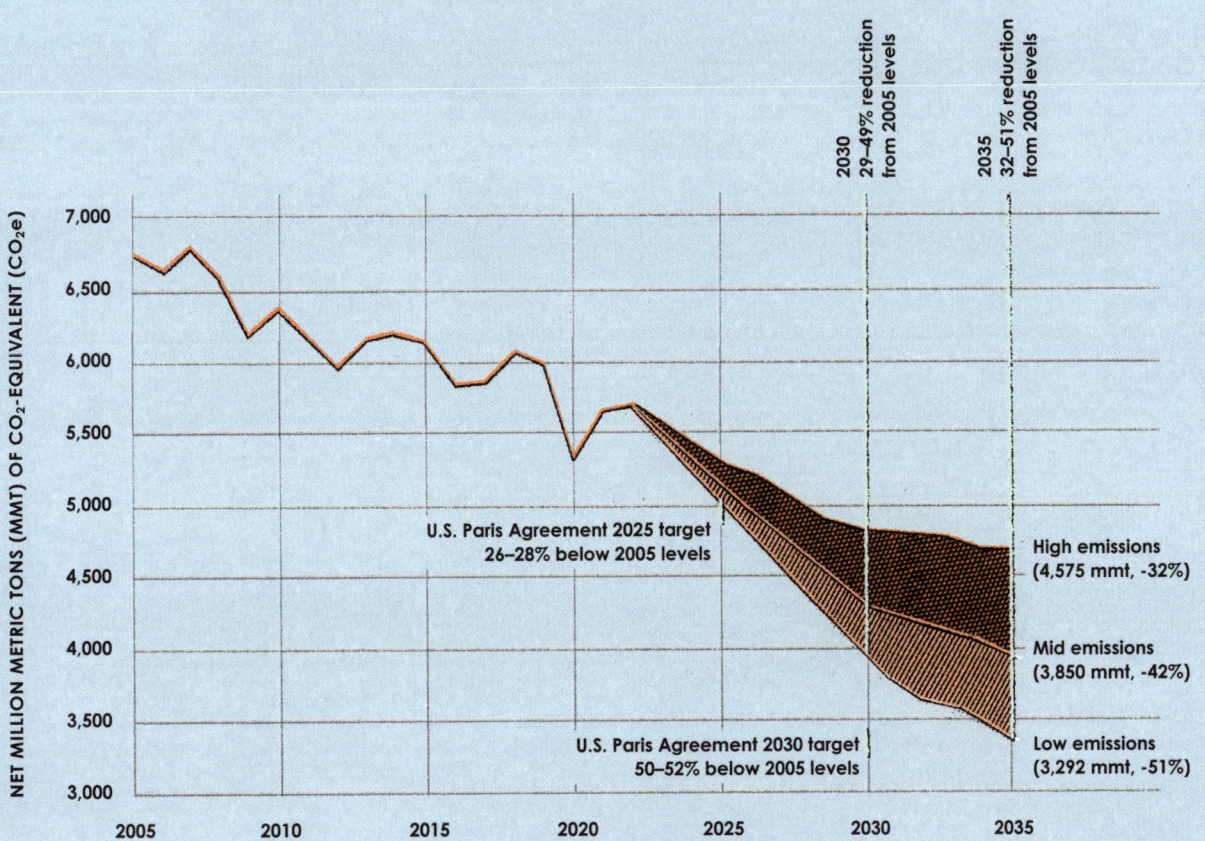

Source: Taking Stock 2023: U.S. Emissions Projections after the Inflation Reduction Act. Rhodium Group, 2023.

1

Manchin, the senior senator from West Virginia, has made millions of dollars off a business that sells a kind of low-grade coal known as "gob." During the 2020 election cycle, he received more donations from the oil-and-gas industry than anyone else in Congress. Manchin strung the White House along for six months before killing off a $2.2-trillion legislative package that, among many other things, would have rewarded utility companies that reduced their emissions and fined those that didn't. Seven months later, he rejected a pared-down version of the plan. "Joe Manchin to Planet Earth: Drop Dead," the headline in *Slate* ran.

The surprise deal came after weeks of secret talks between Manchin and Senate Majority Leader Chuck Schumer. When the two men announced it, the *Politico* headline read "Manchin's Latest Shocker." The bill was predicated on a long list of concessions. It contained tax breaks for putting up wind farms and solar parks, for purchasing electric cars, and for developing energy-storage facilities. It also included tax breaks for aging nuclear plants, and for factories and power stations that continue to burn coal but capture the resulting carbon. It stipulated that, in order for the federal government to lease tracts for offshore wind development, it must lease at least sixty million acres a year for offshore oil and gas exploration. It relied almost entirely on incentives rather than penalties.

The Inflation Reduction Act is the first real piece of climate legislation to make it through Congress. If it does what it's supposed to, it will further reduce the cost of clean energy and spur the growth of whole new industries. Many environmental groups lamented the concessions that had been made to Manchin and the fossil-fuel industry but cheered the bill anyway. The I.R.A. was called "historic," "transformational," and "the best news anyone who cares about the fate of the planet had heard in a very long time."

JOBS, JOBS, JOBS

Eight years ago, Beta and Form didn't exist, and CarbiCrete consisted of four men holding meetings at a Starbucks. Today, more than five hundred people work for Beta, another five hundred work for Form, and fifty work for CarbiCrete. Ørsted's operations in North America employ more than seven hundred people directly and thousands indirectly, through contracts for components, shipping, and logistical support.

Study after study has concluded that cutting emissions creates jobs. Recently, a Princeton-based team issued a report detailing how the United States could reduce its net emissions to zero by 2050. The researchers considered several possible decarbonization "pathways." The one labelled "high electrification" would, they projected, eliminate sixty-two thousand jobs in the coal industry and four hundred thousand in the natural-gas sector. But it was expected to produce nearly eight hundred thousand jobs in construction, more than seven hundred thousand in the solar industry, and more than a million in upgrading the grid.

"For too long, we've failed to use the most important word when it comes to meeting the climate crisis," President Biden declared shortly after taking office. "Jobs, jobs, jobs. For me, when I think climate change, I think jobs."

KILOWATT

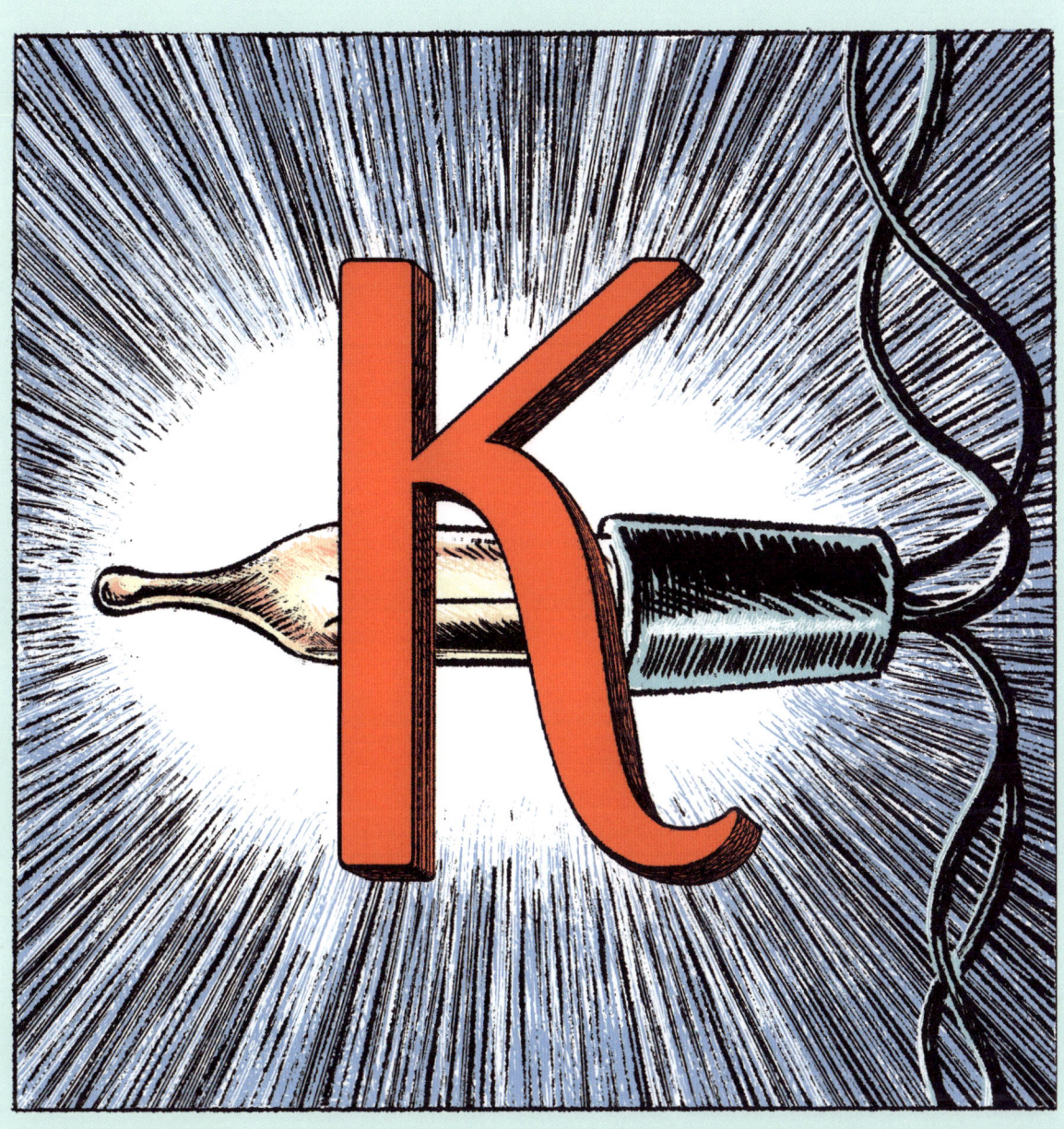

If you add up all the energy America uses in a year—to produce electricity and also to perform the many tasks that have yet to be electrified, like driving and flying and making concrete—and you divide that by the total number of Americans, the result is per-capita consumption. The figure comes to about eighty thousand kilowatt hours. Toss in the energy used to manufacture the goods imported into the United States, and the number rises to almost a hundred thousand kilowatt hours. To put this in terms of power, Americans are consuming roughly eleven thousand watts every moment of every day. A string of incandescent Christmas lights uses about forty watts. It's as if each of us had two hundred and seventy-five of these strings draped around our homes, burning 24/7.

Owing to this every-day-is-Christmas level of consumption, annual emissions in the U.S. run to sixteen metric tons of CO_2 per person. Americans don't have the world's highest per-capita emissions—that dubious honor goes to Kuwaitis and Qataris—but we're up there. Per-capita consumption in Thailand and Argentina runs to around two and a half thousand watts and emissions to around four tons. Ugandans and Ethiopians use a hundred watts and emit a tenth of a ton. Somalis consume a mere thirty watts and emit just ninety pounds. This means that an American household of four is responsible for the same emissions as sixteen Argentineans, six hundred Ugandans, or a Somali village of sixteen hundred.

These figures rarely feature in conversations about climate change in the U.S.; they were hardly mentioned, for instance, in the debate over the Inflation Reduction Act. But to the world's low-consuming countries, the inequities are impossible to ignore. They represent yet another way the Global North has exploited the Global South; call it atmospheric imperialism.

"These disparities chart the rise of developed countries at the expense of others," Mohamed Adow, the director of Power Shift Africa, a Nairobi-based think tank, has written. "The history of climate change is one of compounding injustices."

The average American is responsible for four times the emissions of the average Argentinian, and a hundred and sixty times the emissions of the average Ugandan.

UNITED STATES
American emissions are about 16 tons per capita.

ARGENTINA
Argentinean emissions are about 4 tons per capita.

UGANDA
Ugandan emissions are about 0.1 tons per capita.

LEAPFROGGING

In 1947, the year India gained its independence, telephones were a rarity in the nation; there were fewer than a hundred thousand in the entire country. In the decades that followed, phones remained scarce; as late as 1989, India had just four million for a population of eight hundred and fifty million people. Three-quarters of rural villages lacked any phone connection at all; the official wait time for a line was almost four years, and, when one was finally installed, service was often dismal.

Then, practically all at once, phones were ringing everywhere. In 1994, the country auctioned off its first round of cellular licenses. The auction process was deemed "a mess"; nevertheless, cell service exploded. By 2010, six hundred million Indians were subscribers. (The country's 2011 census revealed that more households had phones than had toilets.) In 2015, cell subscriptions hit a billion. India effectively skipped fixed-line phones and went straight to wireless, a process that's become known as leapfrogging.

Today, India is home to 1.4 billion people. They consume a thousand watts per person, less than one-tenth of what Americans use. Were India to follow the fossil-fuel-slicked development path pursued by China, Europe, and the United States, the result would be planetary disaster. Yet asking India to forgo prosperity on the ground that prosperous nations have already consumed too much is obviously impossible. Fewer than half of all households in the country own a refrigerator. Only one in ten owns a computer. And, even though temperatures in the country's inland cities routinely top a hundred and five degrees Fahrenheit, just one in four has air-conditioning.

L

L

Leapfrogging represents a way—maybe the best way, maybe the only way—out of this dilemma. India is sun-drenched. Instead of building out a grid that relies on coal and natural gas, it could shift to one that relies on solar power and iron-air batteries. Most Indians have never owned a car, so the country could skip over gas-guzzlers and go straight to E.V.s. Ditto for flying. The vast majority of Indians have never been on a plane; the first one they board could be an electric aircraft like the Alice. The same holds true even for stoves. More than five hundred million people in India still cook with wood or dung; instead of transitioning through gas, they could jump straight to induction. In other words, electrify everything!

"India is in a unique position to pioneer a new model for low-carbon, inclusive growth," the International Energy Agency declared in a 2021 report. And what goes for India, the I.E.A. noted, also goes for "a whole group of energy-hungry developing economies."

India "hasn't contributed much to the climate problem," Ashish Gulagi, a researcher at Finland's Lappeenranta University of Technology, told me. "But it can contribute to the solution."

MATH

Carbon dioxide hangs around in the atmosphere for a long time. How long, exactly, is complicated; what matters in terms of the math, though, is not annual but aggregate emissions. This is where the notion of a carbon budget comes from: for every increment of warming, there's a certain amount of CO_2, in total, that can be emitted.

The budget for warming of 1.5 degrees Celsius—almost three degrees Fahrenheit—has, for all intents and purposes, been spent; at current emissions rates, it will run out entirely by 2030. Even the budget for two degrees Celsius—more than 3.5 degrees Fahrenheit—is going fast. It could easily be exhausted within the next few decades.

The United States, with less than five per cent of the globe's population, accounts for twenty-five per cent of aggregate emissions. Europe, with about six per cent of the world's population, is responsible for another twenty per cent. At this point, there's no way to shove all that CO_2 back underground, so no way—or at least no safe way—for the rest of the world to catch up. This ethical challenge is as big as, or perhaps even bigger than, the technical challenge posed by climate change. But let's try to stay positive.

The North grew wealthy by burning fossil fuels. It could use that wealth to help other nations leapfrog to renewables. In 2009, at COP15, in Copenhagen, the world's richest countries took a first step in this direction. They pledged to create a fund to finance clean energy and climate adaptation in countries such as India, Uganda, and Somalia. The fund would grow steadily until, by 2020, it was disbursing a hundred billion dollars a year. Hillary Clinton, who was then the secretary of state, said in Copenhagen that the U.S. recognized the need for "generous financial and technological support," particularly for the "poorest and most vulnerable."

The U.S. is responsible for the largest share of historical emissions, followed by the nations of the European Union.

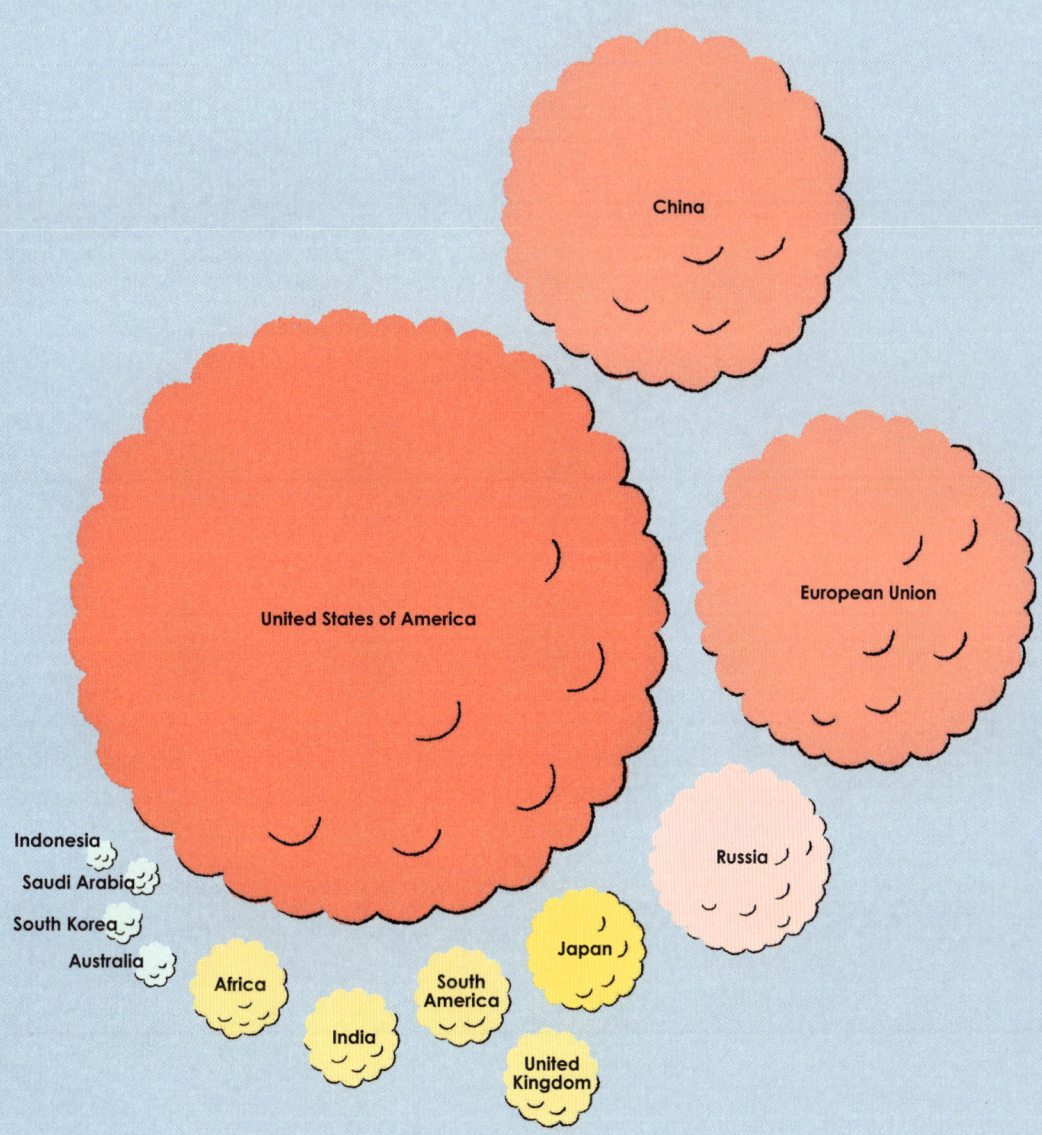

NARRATIVES

"Narratives are socially constructed 'stories' that make sense of events," thereby lending "direction to human action." So observes a paper published recently in the journal *Climatic Change* by a team of European researchers. Climate-change narratives, the team notes, typically foreground "doom and gloom." Often they emphasize risk. If they're not retailing the latest warming-related disasters (fires, floods, food shortages), they're predicting a future filled with even grimmer warming-related disasters (bigger fires, more severe flooding, famines that threaten entire regions).

This approach, the researchers argue, can be counterproductive: "Narratives of fear can become self-fulfilling prophecies." If people believe that things will only get worse, they feel overwhelmed. If they feel overwhelmed, they're apt to throw up their hands, thus guaranteeing that things will only get worse. A diet of bad news leads to paralysis, which yields yet more bad news.

What's needed instead, the paper goes on, are narratives that "empower people to act." Such narratives tell a "positive and engaging story." They "articulate a vision of 'where we want to go'" and outline steps that could be taken to arrive at this metaphorical destination. Positive stories can also become self-fulfilling. People who believe in a brighter future are more likely to put in the effort required to achieve it. When they put in that effort, they make discoveries that hasten progress. Along the way, they build communities that make positive change possible.

N

N

Particularly compelling, by the researchers' account, are "win-win" narratives. Some win-win narratives demonstrate how people can reduce emissions and, in the process, make money. Others center on more intangible goals, such as creating a better world. In a 2008 speech pressing for a "global green new deal," Achim Steiner, then the administrator of the U.N.'s Environment Programme, described the "enormous economic, social, and environmental benefits likely to arise from combatting climate change." One of the key proponents of the Green New Deal in the U.S., Representative Alexandria Ocasio-Cortez, of New York, has argued that a crucial step toward building a more just, more environmentally sustainable future is imagining what this future would look like. "We can be whatever we have the courage to see," she has said.

"Optimism is a choice," notes Christiana Figueres, the Costa Rican diplomat who led the effort to get the Paris climate accord approved.

"Do you know of any challenge in the history of humankind that was actually successful in its achievement that started out with pessimism, that started out with defeatism?" Figueres asked at a conference a few years ago. "There isn't one," she said, answering her own question.

OBJECTIONS

"The gap between wishful thinking and reality is vast." So observes Vaclav Smil, a professor emeritus at the University of Manitoba. The observation could apply to almost anything; Smil, who has written more than a dozen books about energy and society, is concerned with the gap between the aspiration to fight climate change and the immense on-the-ground effort entailed in actually doing so. Studies that purport to show how the world could radically reduce or eliminate its carbon emissions by one date or another tend, he argues, to presuppose what they claim to be proving. To arrive at their foregone conclusions, they rely on a variety of unreliable assumptions—that existing technologies will be deployed at fantastic rates, or that nonexistent technologies will be deployed at fantastic rates, or that humanity's ever-growing appetite for energy will suddenly be curbed, or some combination of all three. Smil labels such studies "the academic equivalents of science fiction."

Everything I have written, from "Despair" onward, is vulnerable to Smilian objections. Consider "Flight." It's possible that, in a few years, Alias ferrying pallets of cargo will zip between regional airports. It's also possible that electric passenger planes will one day make short hops between, say, Boston and Hyannis. But that could be the limit. The world's best-selling passenger plane, the Boeing 737, can transport some two hundred people coast to coast. To electrify such a flight would require more than eight hundred tons' worth of current-generation lithium-ion batteries, or four hundred tons of lithium-ion batteries functioning at their maximum theoretical capacity. To get off the runway, though, a 737 can't weigh more than eighty tons, passengers and crew included. A recent paper by researchers at Carnegie Mellon concluded that the demands of larger aircraft lie beyond the "feasibility limits" of known battery technologies.

As emissions continue, it becomes that much more difficult to limit warming to 1.5 degrees Celsius. At this point, to reach the target, emissions would have to be eliminated entirely within the next decade.

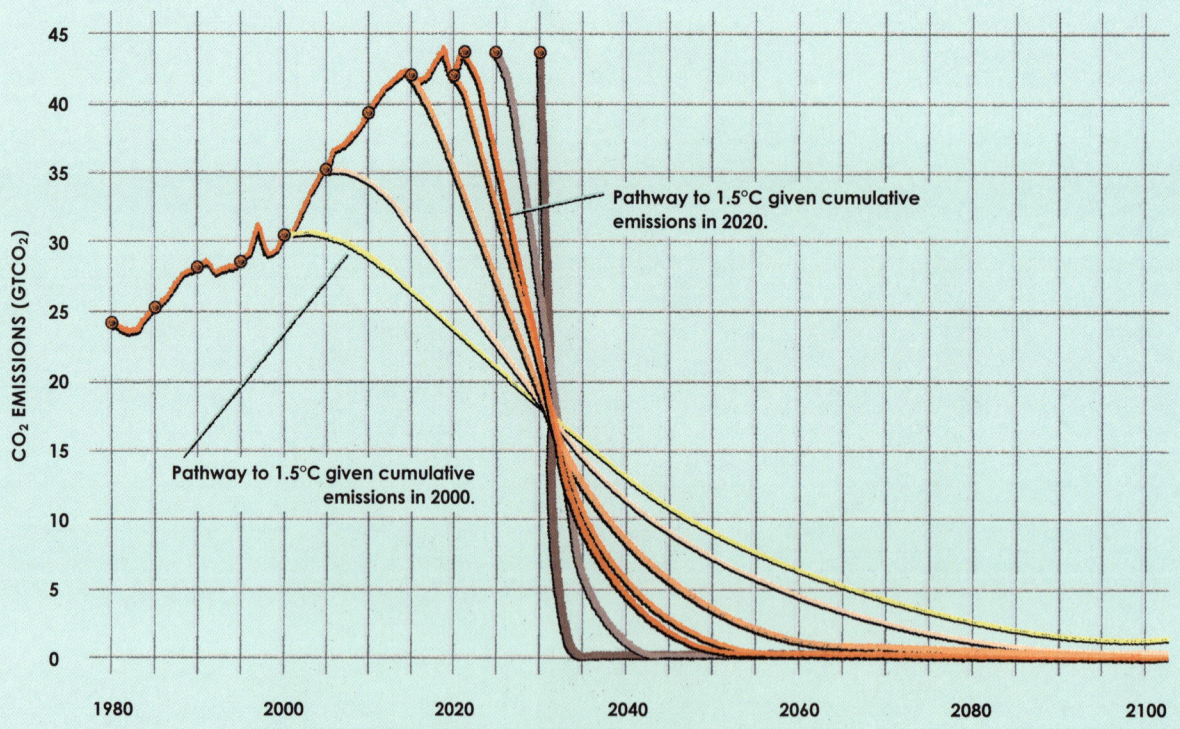

Source: CarbonBrief

Or consider "Green Concrete." As promising as CarbiCrete may be, the niche it fills, much like the Alia's, is a narrow one. Since it has to be cured in chambers filled with concentrated CO_2, CarbiCrete can't be poured at a work site; it can be used only for pre-cast products, like cinder blocks or patio tiles. Meanwhile, though the blocks and tiles absorb CO_2 as they harden, a great deal of CO_2 was released in the process of producing the slag that went into them; globally, the steel industry is responsible for roughly the same number of tons of emissions as the concrete industry—around three billion.

To say that amazing work is being done to combat climate change and to say that almost no progress has been made is not a contradiction; it's a simple statement of fact. At the time of the Rio summit, fossil fuels provided roughly eighty per cent of the world's primary energy. Today, fossil fuels still provide nearly eighty per cent of the world's primary energy. In the meantime, total global energy use has increased by almost two-thirds. As Smil puts it, "The inertia of large, complex systems is due to their basic energetic and material demands—as well as the scale of their operations."

POWER

The United States's power grid has been called "the largest machine ever built by man." It comprises more than eleven thousand generating plants, more than six hundred thousand miles of high-voltage transmission lines, and some six million miles of distribution lines. Several recent studies claim to show that decarbonizing the grid in the nearish future is feasible. All of them, as per Smil, involve a certain amount of "science fiction"; they describe what is technically possible while glossing over the potential barriers to implementation. These barriers, though, are myriad. Some are economic, some are legal, some are logistical, some are political, and some, to paraphrase Polonius, are legal-logistical or economic-legal-logistical-political.

QUAGMIRE

Take what's been called the "transmission quagmire." To clean up America's grid, it's not enough to build new generating capacity, or even new generating capacity plus new storage capacity. Power has to be transported from places that have a lot of wind and sun to urban centers that use a lot of electricity. Decarbonizing the grid will, by one estimate, demand more than a million miles of new transmission lines, and the cost of stringing all these lines will, by another estimate, come to more than two trillion dollars. Managing such a gargantuan project would be difficult enough if someone were in charge. But thanks to the way the grid was put together—bit by bit, over many decades—jurisdiction over transmission lines is divided among an electoral map's worth of competing authorities.

Whenever lines cross state borders, the mire becomes particularly quaggy. In that case, each state's utility commission has to sign off. In some states, every affected county does, too. Then there are the local utility companies, which may, officially or unofficially, hold veto power.

"Let's say I'm a local utility, and you tell me all this low-cost power is going to come in from out of state with a new transmission line," Steve Cicala, an economics professor at Tufts, said to me. "My reaction is 'Absolutely not. It's a threat to my business model.' And a lot of public-utility commissions are pretty much captured by the local utilities."

Q

The seven-hundred-and-twenty-mile Plains and Eastern Clean Line was supposed to link wind farms in Oklahoma to customers in Tennessee; it was killed by opposition from Arkansas. The Grain Belt Express was designed to run from southwest Kansas to Indiana; it's been delayed for a decade, thanks to resistance from Missouri. The TransWest Express is intended to bring wind power from Wyoming to cities on the West Coast; construction has been held up for years, in good part owing to a single litigious family in Colorado.

Northern Pass was a transmission line designed to bring hydropower from Quebec to Massachusetts via New Hampshire. After New Hampshire rejected the project, in 2018, Massachusetts announced that it would try going in a different direction. It would build a line, dubbed New England Clean Energy Connect, that would cut through Maine instead. Work on NECEC was already underway when, in the fall of 2021, Maine voters approved a referendum to block its completion. In the spring of 2023, a Maine jury ruled that the referendum was void and that work on the line could continue. The delay has increased the cost of the project by at least half a billion dollars.

REPUBLICANS

Reaching net zero in the United States will require putting such wrangling aside. It will require building out the transmission system while, at the same time, expanding its capacity so that hundreds of millions of cars, trucks, and buses can be run on electricity. It will require installing tens of millions of public charging stations on city streets and even more charging stations in private garages. Assembling the electric cars and trucks will, in turn, necessitate extracting nickel and lithium for their batteries, which will mean siting new mines, either in the U.S. or abroad. The new cars and trucks will themselves have to be manufactured in an emissions-free manner, which will involve inventing new methods for producing steel or building a new infrastructure for capturing and sequestering carbon, or both.

The list goes on and on. The fossil-fuel industry will essentially have to be dismantled, and millions of leaky and abandoned wells sealed. Concrete production will have to be reengineered. The same goes for the plastics and chemicals industries. Currently ammonia, a critical component of fertilizer, is produced from natural gas, so the fertilizer industry will also have to be refashioned. Practically all the boilers and water heaters that now run on oil or gas, commercial and residential, will have to be replaced. So will all the gas stoves and dryers and industrial kilns. The airline industry will have to be revamped, as will the shipping industry. Farming is responsible for roughly ten per cent of America's greenhouse-gas emissions, mostly in the form of nitrous oxide and methane. (Nitrous oxide is a byproduct of fertilizer use; methane is released by rotting manure and burping cows.) Somehow, these emissions, too, will have to be eliminated.

All of this should be done—indeed, *must* be done. Officially, the U.S. is committed to reaching net zero by 2050. But a task of this scale has never been attempted before. Zeroing out emissions means rebuilding the U.S. economy from the bottom up. Perhaps Americans recognize this, perhaps not. In the summer of 2022, at a time when much of the country was baking in ninety-five-degree-plus heat, the *New York Times* took a poll of registered voters. Asked to name the most important problem facing the nation, twenty per cent of the respondents said the economy, fifteen per cent said inflation, and eleven per cent said partisan divisions. Only one per cent said climate change. Among registered Republicans, the figure was zero per cent.

R

SHORTFALL

And what goes for the United States goes for the rest of the world. China is currently the world's biggest emitter on an annual basis. It has said that it will reach net zero by 2060. In 2022, Beijing approved new coal-fired generating plants at the rate of two a week. Peak electricity demand in China has been rising rapidly owing to ever-more-punishing heatwaves, a trend that will only be exacerbated by more coal burning. Climate Action Tracker, an independent research group based in Berlin, rates China's climate policies as "highly insufficient" and notes that the country has recently "reneged" on promises to shift away from coal.

India, which is now the world's third-biggest emitter—one rung behind the U.S.—has declared that it will reach net zero by 2070. In the meantime, its emissions are rising fast. The country still relies on coal for more than seventy per cent of its electricity production, and in recent years the government has reopened coal mines that had previously been deemed unprofitable. "Our energy needs are first and foremost," India's coal secretary, Amrit Lal Meena, told the *Washington Post*.

Russia is the world's fourth-largest emitter. It is aiming, on paper at least, to reach net zero by 2060. But it has taken almost no action to rein in its emissions, and this seems unlikely to change anytime soon. In 2003, President Vladimir Putin joked that warming would enable Russians to "spend less on fur coats"; more recently, he has blamed climate change on "some invisible shifts in the galaxy."

The European Union's pledge to hit net-zero emissions by 2050 is written into E.U. law. But, after Russia cut gas deliveries to the bloc, several countries, including Germany and the Netherlands, announced plans to fire up old coal plants or extend the lives of plants that had been slated to close. "The war in Ukraine is putting climate action on the back burner," the U.N. secretary-general, António Guterres, lamented. A recent study by a consortium of research institutes in Europe and the U.S. concluded that only five per cent of the hundred and twenty-eight countries that have set the goal of reaching net zero have taken the requisite first steps. It found an "alarming lack of credibility" to most of the commitments.

The broken promises extend in all directions. At COP26, held in Glasgow in 2021, a hundred and forty countries vowed to put a stop to "forest loss," another major source of greenhouse-gas emissions. (A single full-grown tree can store as much as seven tons of carbon.) Among the signatories of the declaration was Brazil; in the six months following the conference, deforestation rates in the country soared to new highs.

The Democratic Republic of the Congo also signed the forest pledge. In the summer of 2022, the D.R.C. announced that it was opening thirty huge tracts, encompassing an estimated twenty-five million acres of rain forest and two million acres of peatlands, to oil and gas exploration. (Some of the tracts overlap Virunga National Park, a refuge for critically endangered mountain gorillas.)

Then there are the pledges of financial support. The hundred billion dollars a year promised in Copenhagen to developing nations have yet to materialize. By some estimates, the shortfall runs to twenty billion dollars a year; by others, it's four times that amount.

"To be charitable, this is a manifestation of the inability of the developed countries to deliver," Saleemul Huq, the director of the International Centre for Climate Change and Development, in Dhaka, Bangladesh, told me. "To be uncharitable, they just don't give a damn. Unfortunately, the uncharitable interpretation seems to be more valid."

TEMPERATURES

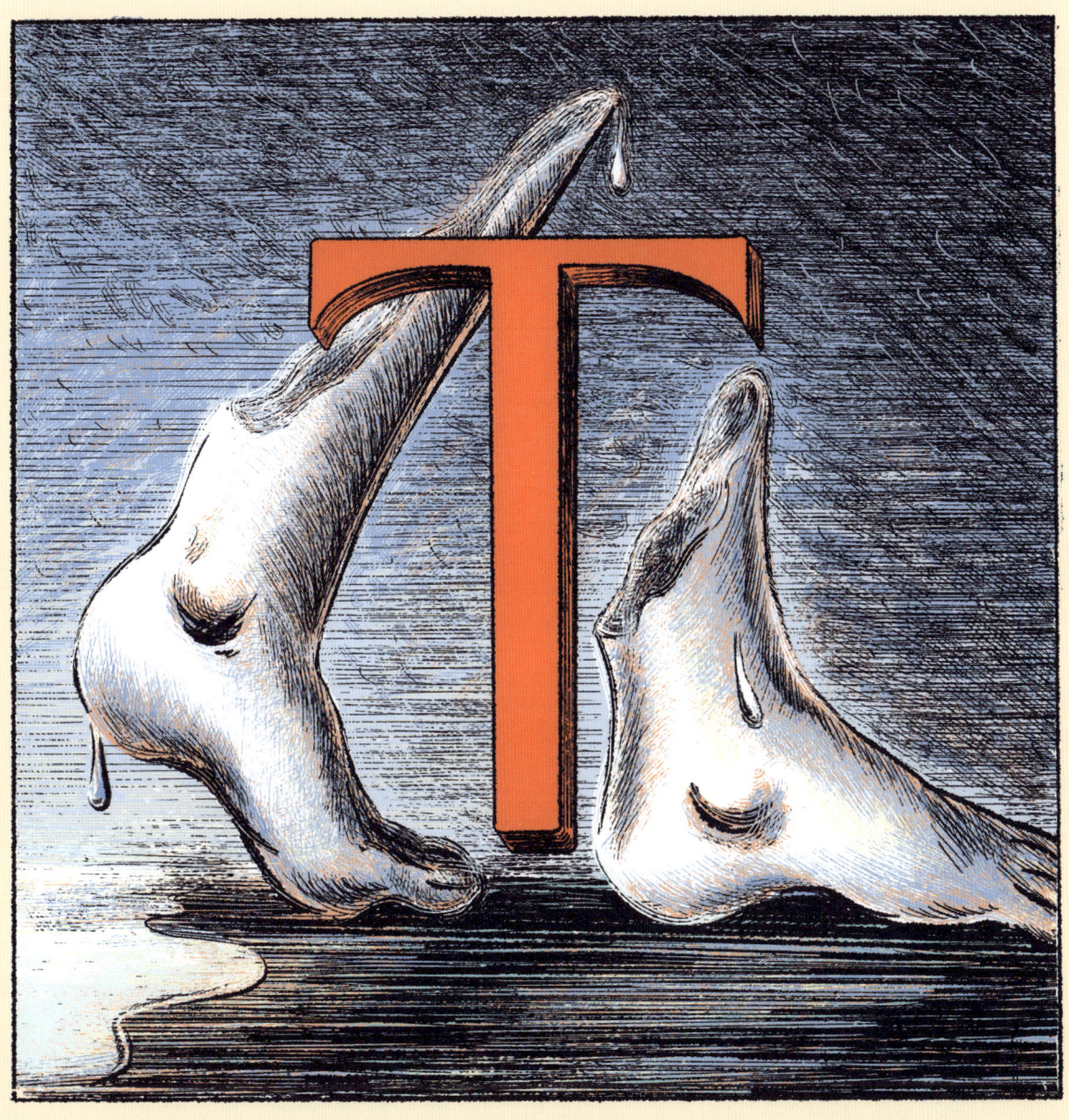

The Institute for Exercise and Environmental Medicine, in Dallas, offers a hyperbaric chamber where divers can recover from the bends, a pool equipped to continuously measure swimmers' oxygen use, and a climate-controlled vault that can be programmed to test the limits of human endurance. Not long ago, I swallowed a thermometer the size of a pill and had myself sealed in the vault.

Formally known as the environmental chamber, the vault resembled a walk-in freezer, with metal walls and a pressed-metal floor. Pretty much every available surface was occupied by machinery—computer screens, thermocouples, an electrocardiogram monitor, a treadmill, and a sort of stationary bicycle that looked like a suitcase with pedals. In the center sat a lawn chair, which a technician indicated I should take.

With me in the chamber was a researcher named Josh Foster. Before he allowed me to enter the vault, Foster had asked for a urine sample—a first in my reporting career. He'd also stuck some electrodes on my chest and performed an ultrasound scan of my heart, which, he said, was unusually low and hard to find.

Foster, who is British, is interested in the effects of extreme heat on the body. To this end, he creates miniature heat waves and solicits volunteers to sweat their way through them. On the day I volunteered, the temperature in the vault was a hundred and six degrees Fahrenheit and the humidity forty per cent.

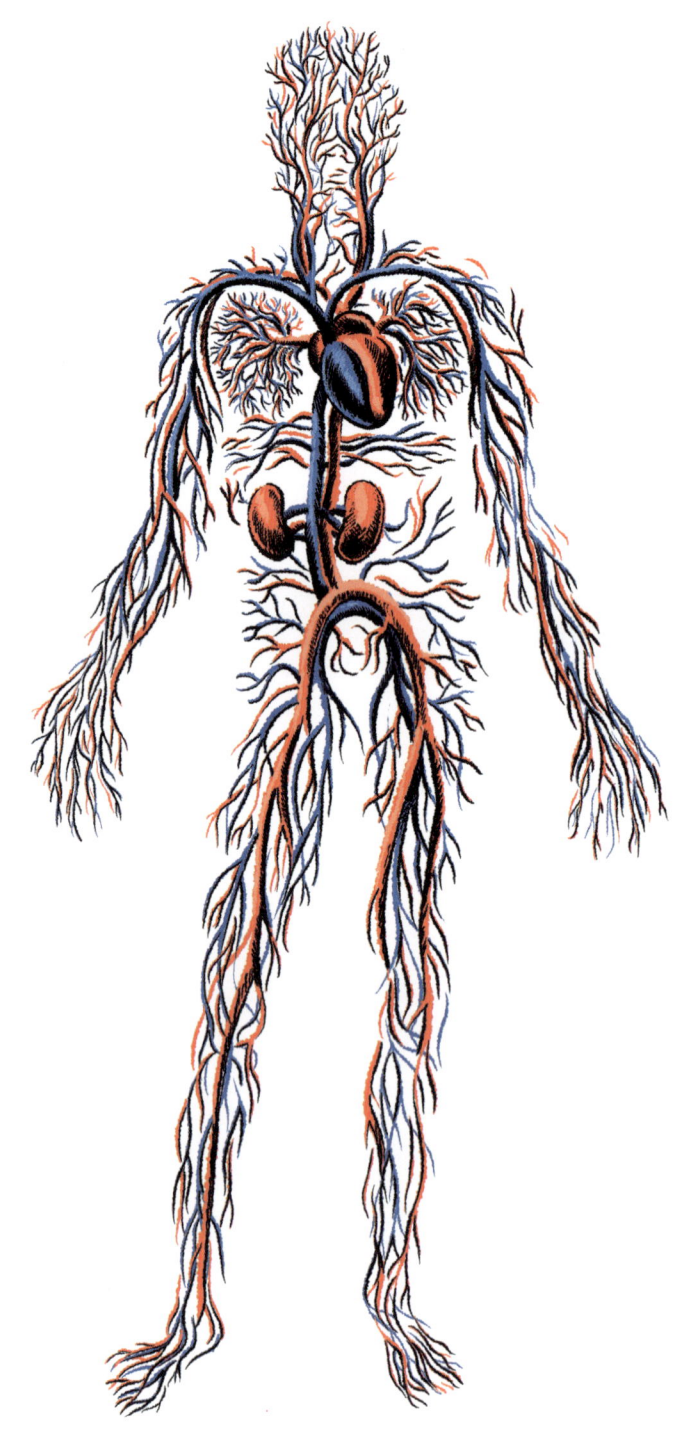

"Temperature regulation is one of the most important variables the body will try to protect," Foster told me. "Because as soon as you start to stray from what's normal, outside of a given quite small range, our ability to tolerate that is very, very low."

Once a topic of marginal academic interest, the physiology of heat stress is now a subject of widespread practical concern. According to a recent study, two hundred and seventy-five million people around the globe are subjected to life-threatening temperatures at least one day a year, and this number could easily grow to eight hundred million by the middle of the century. According to another recent study, the incidence of "extreme humid heat" has doubled in the course of the past forty years. Some parts of the world, particularly in South Asia and around the Persian Gulf, are already experiencing temperatures close to the human "survivability limit." In the summer of 2023, heat record after heat record fell. In the city of Ahvaz, in western Iran, the mercury hit a hundred and twenty-three degrees Fahrenheit, and in Sanbao, in northwestern China, it reached hundred and twenty-six degrees. Phoenix saw a hundred and eighteen degrees, Tucson a hundred and twelve, El Paso a hundred and eleven, and Austin a hundred and seven.

The human body reacts to such temperatures by sweating and directing more blood toward the skin. Problems arise when people become dehydrated, or their hearts get overtaxed, or it's just so sweltering that they can't dissipate enough heat. The elderly are particularly vulnerable to heat stress, Foster told me, because they sweat less than young people, and their hearts don't pump as efficiently. (Humidity impedes the evaporation of sweat, which is why extreme humid heat is so dangerous.) One consequence of prolonged heat exposure can be a kind of blood poisoning.

"Increased blood flow to the skin means that less blood is being directed toward the gut," Foster explained. "And, if that happens for a long enough time, it can damage the cells that line the gut, and bacteria that are normally housed in the gut can leak out. It's basically the same as having sepsis." The heat wave that affected most of Europe in the summer of 2022 is estimated to have killed more than sixty thousand people.

Sitting in the environmental chamber, with the pill-size thermometer in my stomach, would, I hoped, be edifying without being too edifying. Until the United States and the other big emitters reach net zero—indeed, until the entire world reaches net zero—the planet will continue to warm. What is the future we're creating actually going to feel like?

Every quarter of an hour, I was supposed to ride the stationary bicycle for five minutes; this was to simulate the sort of effort a person would have to make in the course of completing ordinary household chores. I started off strong but after a few rounds began to flag. The humidity made the air seem strangely solid. I tried to imagine what it would be like to perform real work under these conditions but found it difficult to hold on to a thought.

A few days later, when I got back home, Foster sent me the data that had been collected by the various instruments. I had sweated out almost a pint of water every hour. My heart rate had increased by thirty beats a minute and the blood flow through my brachial artery had more than tripled. Despite all the (admittedly involuntary) effort I had made to thermoregulate, my core temperature had risen to a hundred degrees.

T

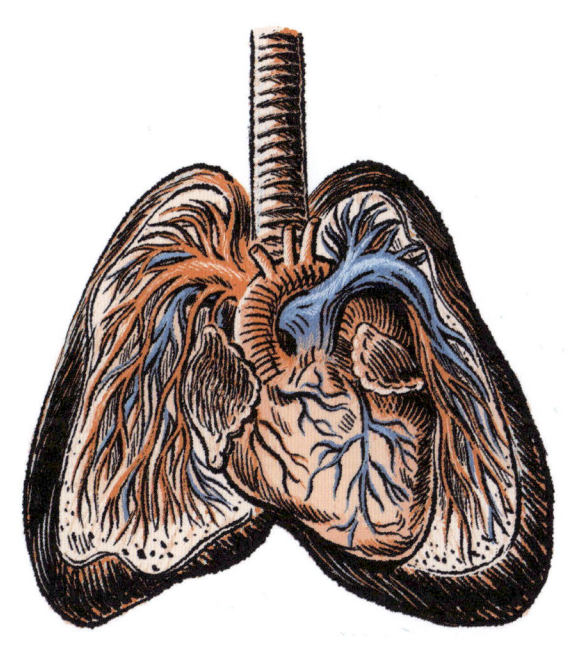

UNCERTAINTY

Over the past billion years, the Earth's temperature has fluctuated wildly. Around seven hundred million years ago, in the period known as the Cryogenian, the entire planet was covered with ice. "Snowball Earth" thawed, only to be plunged into another global glaciation. About ninety million years ago, during what's known as the Cretaceous Thermal Maximum, breadfruit trees grew in northern Greenland and the tropical oceans were as toasty as a hot bath. In our own period, the Quaternary, the swings have been spectacular; at least twenty times in the last two and a half million years, glaciers have pushed south from the Arctic and then retreated again. The ice ages themselves were marked by dramatic temperature oscillations. The last one, which ended about twelve thousand years ago, went out, in the words of one glaciologist, in a "drunken stagger."

You can't prepare for a future you can't imagine. The trouble is, it's hard to picture the future we are creating. As the climate swings of the past suggest, even subtle and gradual forces—tiny variations in the Earth's orbit, for example—can have world-altering consequences. And what we're doing now is neither subtle nor gradual. In little more than a century, humans have burned through coal and oil deposits that took tens of millions of years to create.

Climate change is characterized not just by uncertainty but by something risk analysts call "deep uncertainty." There are known unknowns to worry about, and unknown unknowns.

U

VAST

Climate surprises keep popping up. Starting in 2007, for example, methane levels in the atmosphere took an unexpected jump. Methane is a far more potent greenhouse gas than CO_2, so scientists were alarmed. They eventually figured out, on the basis of the methane's isotopic composition, that the source of the increase couldn't be fossil-fuel production, even though oil and gas wells often leak methane into the air. Instead, the culprit must be microbes, either the sort that live in a marsh or the sort that live in a cow's gut. Recent research suggests that the bulk of the extra methane is coming from the Sudd, a huge wetland in South Sudan, and that warming itself is responsible for the uptick in microbial activity. If that's the case, then a spiral is likely to ensue: more methane will produce more warming, which will produce yet more methane, and so on.

V

V

How many positive feedback loops like this have already been—or are about to be—initiated? Despite the best efforts of climate modelers, no one can say. Several enormous Antarctic glaciers rest on bedrock that's below sea level; as these glaciers retreat, water is starting to seep underneath and to melt them from the bottom up. This, in turn, is leading to more retreat and still more melting. One retreating glacier, formally known as Thwaites, has informally become known as the Doomsday Glacier. A recent paper in *Science* observed that the "eventual collapse" of Thwaites, which is the size of Florida, "may already be inevitable." Even after global emissions reach net zero—whenever that is—ice sheets will continue to melt and sea levels to rise for hundreds or perhaps thousands of years.

All the way back in 1965, the authors of one of the first reports on global warming, which was not yet known as global warming, warned that humanity was "unwittingly conducting a vast geophysical experiment." As Marcia Bjornerud, a geologist at Lawrence University, has written, the irony of our oversized impact on the Earth is that we have "put Nature firmly back in charge, with a still-unpublished set of rules we will simply have to guess at."

WEATHER

The National Oceanic and Atmospheric Administration tracks weather-related disasters in the United States that cause more than a billion dollars' worth of damage. According to NOAA, in the nineteen-eighties the U.S. saw an average of three such disasters per year. In the nineteen-nineties, the average was five per year; in the two-thousands, it was six; and in the twenty-tens it jumped to twelve. (The figures have been adjusted for inflation.) In 2020, a record-shattering twenty-two disasters costing more than a billion dollars struck the country. This record was almost matched in 2021, with twenty billion-dollar disasters, and again in 2022, with eighteen. Then, in 2023, the record was smashed again: in the first eight months of the year, there were twenty-three weather-related disasters that caused more than a billion dollars' worth of damage. Adam B. Smith, a NOAA researcher, has written that a disastrous number of disasters "is becoming the new normal." The rise is partly a function of more people living in vulnerable areas, such as floodplains. But increasingly it's a function of climate change.

W

In the future, the costs may climb steeply or they may climb precipitously. All our infrastructure has been built with the climate of the past in mind. Much of it will have to be rebuilt and then, as the world continues to warm, rebuilt again.

To protect the Houston area (and its many petrochemical plants) from rising seas and storm surges, the U.S. Army Corps of Engineers is planning to erect a huge system of gates at the mouth of Galveston Bay. The price tag for the project, known as the Ike Dike, is estimated at thirty billion dollars. Norfolk, Virginia, is hoping to stave off the water with a $1.5-billion series of barriers, levees, and tidal gates, and Charleston, South Carolina, is looking to build a billion-dollar floodwall. Some places—large swaths of Miami, for instance—may prove impossible to defend, meaning that real estate now valued in the hundreds of billions of dollars will have to be written off.

XENOPHOBIA

One of climate change's many compounding injustices is that the highest costs will be borne by those who have contributed the least to the problem. Several low-lying island nations, including Tuvalu and Kiribati, are destined simply to disappear. In Bangladesh, some two thousand people arrive every day in the capital, Dhaka, many driven by storms or rising seas that have made village life unbearable. In Pakistan, in 2022, flooding caused by supercharged monsoon rains killed a thousand people and forced six hundred thousand more into relief camps.

The United Nations High Commissioner for Refugees has estimated that, globally, an average of twenty-one million people are being displaced by weather-related events every year. The U.N.'s International Organization for Migration has projected that by 2050 as many as a billion people may be on the move. In the coming decades, "huge populations will need to seek new homes," Gaia Vince, a British journalist, has written. Either "you will be among them, or you will be receiving them."

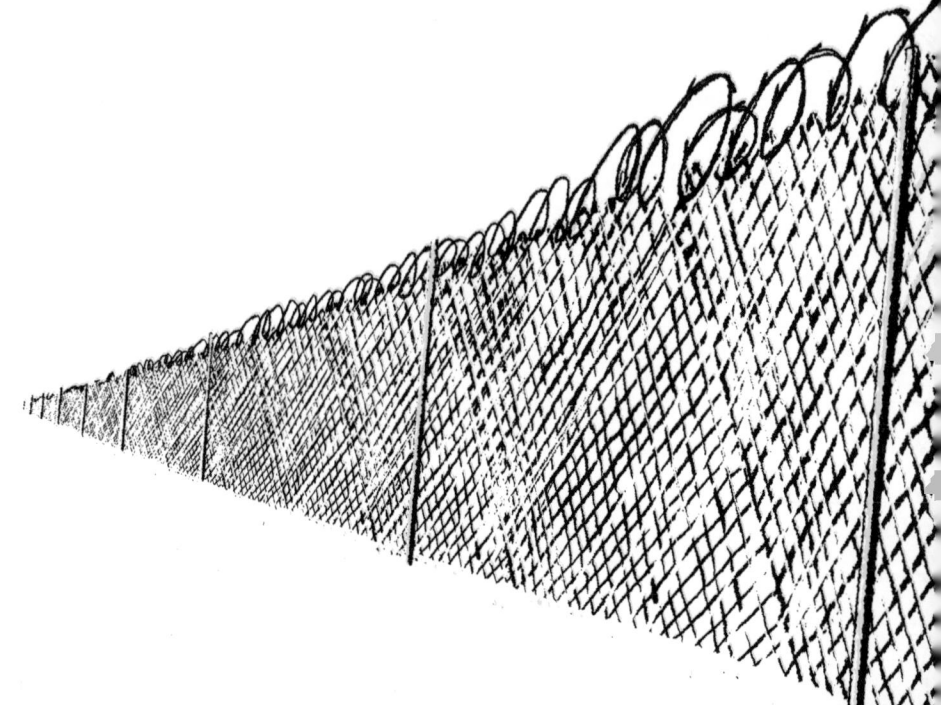

X

X

Almost as much as climate change itself, this great displacement will test national and international institutions. One possibility is that climate refugees will be welcomed. This could happen because it's the right thing to do, or it could happen for less high-minded reasons. As Akka Rimon, a former foreign secretary of Kiribati, has observed, "Countries like Australia need workers," while the citizens of countries such as Kiribati will soon need a different place to live. These needs are complementary. The European Union, too, faces a labor shortage. A communiqué issued recently by the European Commission noted that there's a strong "economic case" for allowing in more legal immigrants, especially since "the transition to a climate-neutral economy" will require "additional labour and new skills." Climate migrants could play a key role in decarbonization, providing a new kind of win-win narrative.

Another possibility is that climate migrants, like millions of migrants before them, will be despised. Rich countries—including those in the E.U.—will try to keep refugees out, and those who manage to slip in will be herded into camps. In an effort to gain power, right-wing politicians will vilify them, and this will encourage yet more racism and xenophobia—a social feedback loop. Giorgia Meloni, Italy's prime minister, has said that her country ought to "repatriate the migrants back to their countries, and then sink the boats that rescued them."

Both the effort to limit climate change (by replacing the world's energy systems) and the effort to adapt to climate change (by erecting dikes and seawalls) will take place in the context of climate change, which is to say as cyclones, drought, fire, and sea-level rise force millions of people to flee. It's possible that cascading crises will accomplish what three decades of climate negotiations have not, and unite the world to seek the best way forward. Or it's possible that the same forces that have prevented cooperation in the past—nationalism, corporatism, sectarianism, fear—will, under the stress of climate change, only intensify.

YOU

So far, average global temperatures have risen by 1.1 degrees Celsius—two degrees Fahrenheit—and the budget for 1.5 Celsius is nearly gone. How hot will it get? Will temperatures climb two degrees Celsius? Two and a half? Three? A study published a few years ago, by Veerabhadran Ramanathan, a climate scientist at the Scripps Institution, and Yangyang Xu, of Texas A&M, defined a temperature increase of 1.5 degrees as "dangerous," an increase of three degrees as "catastrophic," and an increase of five degrees as "unknown, implying beyond catastrophic." A second study, by a group of American and European researchers, determined that, if we were to burn through all known fossil-fuel reserves, global temperatures could rise by as much as eleven degrees Celsius, or twenty degrees Fahrenheit. (How humanity could keep the oil flowing even as the world drowned and smoldered was a question the researchers did not address.)

Y

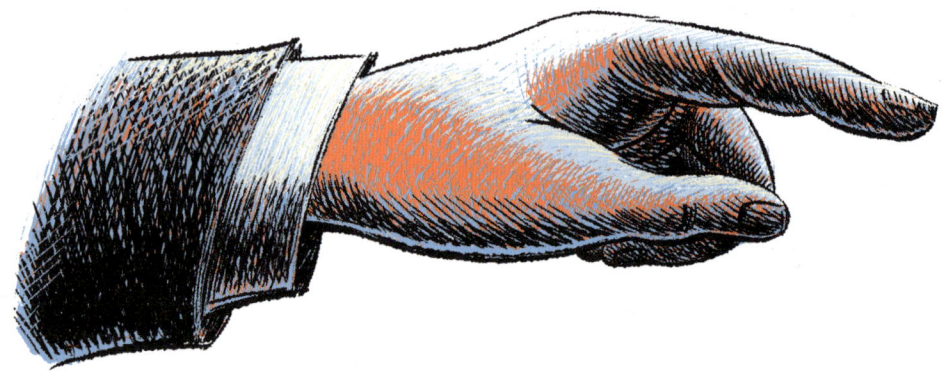

Y

There are good reasons to opt for optimism. (See "Narratives.") It could be argued that the passage of the Inflation Reduction Act was possible only because so many people believed in a better future. At the same time, there are good reasons to wonder whether optimism lies at the heart of the problem. For the last thirty years—more if you go back to 1965—we have lived as if someone, or some technology, were going to rescue us from ourselves. We are still living that way now.

"You can't just sit around waiting for hope to come," Greta Thunberg observed in a speech scolding E.U. politicians. "Then you're acting like spoiled, irresponsible children. You don't seem to understand that hope is something you have to earn."

ZERO

In the summer of 2022, I rented a car in Las Vegas and drove out to Hoover Dam. There I signed up for a tour that began with an educational video. Construction of the dam, the narrator of the video intoned while grainy black-and-white footage jittered across the screen, entailed pouring more than three million cubic yards of concrete. Put to a different use, this much concrete could pave "a four-foot-wide sidewalk around the earth's equator." When the dam was completed, in the middle of the Depression, it "gave new life to the desert Southwest" as well as "to the nation's spirit." The Colorado River began backing up behind the massive structure to form Lake Mead, the country's largest reservoir, which can store enough water to "cover the entire state of Pennsylvania to the depth of one foot."

After the video, my tour group took an elevator down thirty stories, into the dam's hydroelectric plant. Here we were regaled with more facts: Hoover Dam is equipped with seventeen generators—eight on the Nevada side of the river and nine across the border, in Arizona. Each generator can produce enough electricity—a hundred and thirty megawatts—to power sixty-five thousand homes. Each contains five miles' worth of copper wire and a hundred and sixty tons' worth of electromagnets. The tour ended on an observation deck where an audiotape of yet more dam-related facts—the structure weighs 6.6 million tons and is twelve hundred and forty-four feet long—was issuing from a loudspeaker. The narrator of the audiotape sounded an awful lot like the narrator of the video. "It has been said that in the shadow of Hoover Dam one feels that the future is limitless, that we have in our grasp the power to achieve anything, if we can but summon the will," he concluded. Then the tape started over.

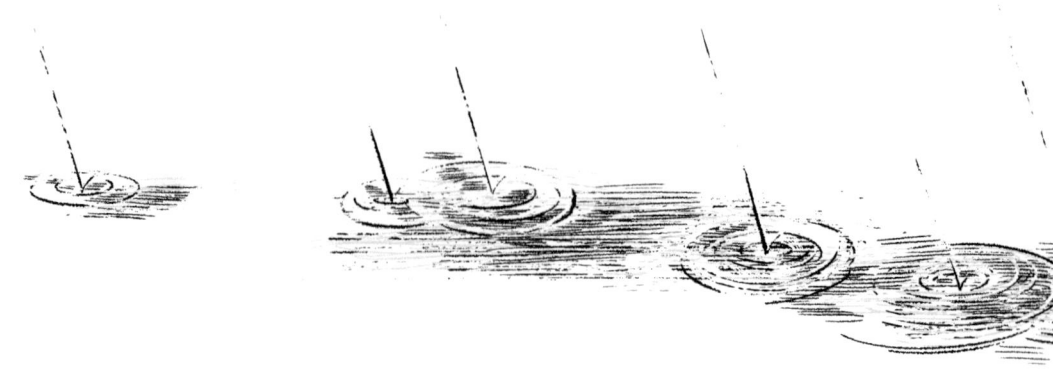

The Colorado River basin has been called "ground zero for climate change in the United States." If this is the case, then Hoover Dam might be described as ground zero's ground zero. Since 1998, the basin has been stuck in a drought; this drought has lasted so long and grown so deep that, by the time of my visit, it was being referred to as a "megadrought."

From the observation deck, the drought's effects were scarily apparent. An abandoned dock lay, in pieces, high above the lake's edge. Instead of being submerged, the power plant's four intake towers stuck up into the air, like lighthouses. The steep walls of the reservoir, which in pre-dam days formed Black Canyon, were lined with an enormous white stripe—a geological oddity known as the bathtub ring. The ring, composed of minerals deposited by the retreating water, runs as straight as a ruler, mile after mile. At the start of the drought, the stripe was as high as a giraffe. By 2015, it had grown as tall as the Statue of Liberty. In 2022, it reached the height of the Tower of Pisa. The water level was so low that the dam's seventeen generators could operate only sporadically.

I had wanted to talk about the dam, the megadrought, and the future of the Colorado River basin with a representative of the Bureau of Reclamation, which built and still operates Hoover Dam. But when I got in touch with the bureau's office in Boulder City, Nevada, a town created to house the workers who erected the dam, I was told that no one there was giving interviews. I was, of course, welcome to take a public tour. I ended up taking two. On the first, no mention was made of the drought; on the second, I tried to force the issue. I asked the guide whether she got any questions about Lake Mead, which at that time was only about a quarter full. She said she did, but she wasn't supposed to answer them. "We're not to comment too much on it," she told me.

"You know, I haven't been on the lake at all this year," she added. "It's just sad when I go out there. It's a little depressing. To save my sanity, I don't go." Lake Mead used to be lined with boat launches; most of these are now closed.

The construction of Hoover Dam was authorized in 1928, just a year after Svante Arrhenius died. The project reflects the same faith in progress that he held to—a faith in humanity's power to improve on nature. This is still the faith that the Bureau of Reclamation is pushing even as the logic of Lake Mead comes undone. In the winter of 2023, parts of the Colorado River basin experienced near-record snowfall; even with all the runoff, the reservoir remained in what's technically known as a Level 1 Shortage. The wet winter was followed by a phenomenally hot summer that pushed much of the Southwest, including the area around the lake, back into drought conditions.

Following my second tour of the dam, I climbed up to the observation deck again to take a last look around. It was almost noon, and the desert sun was high overhead. A couple of tour groups came and went as the tape played in the background: "In the shadow of Hoover Dam, one feels that the future is limitless . . . limitless . . . limitless." What I felt standing in the dam's shadow was something different.

Climate change isn't a problem that can be solved by summoning the "will." It isn't a problem that can be "fixed" or "conquered," though these words are often used. It isn't going to have a happy ending, or a win-win ending, or, on a human timescale, any ending at all. Whatever we might want to believe about our future, there are limits, and we are up against them.

FROM ELIZABETH KOLBERT

I had a lot of help on this project. First, I'd like to thank all the people who shared their expertise with me. Many are named in the previous pages; many others are not; I am equally grateful to you all. Much of this text originally ran in *The New Yorker*. My editors at the magazine—David Remnick, Daniel Zalewski, and Carla Blumenkranz—offered invaluable advice while it was coming together, and Anya Kordunsky went beyond the call of duty to ensure its accuracy. Special thanks to the team at Ten Speed Press, who turned the text into such a beautiful book, and in particular to Molly Birnbaum and Betsy Stromberg, who truly were a pleasure to work with. I feel very fortunate to have been able to collaborate on this project with Wesley Allsbrook, whose talent and creativity shine through in every illustration. As ever, I am indebted to my agent, Kathy Robbins, and to my husband, John Kleiner.

FROM WESLEY ALLSBROOK

I owe great thanks to Alexandra Zsigmond and Aviva Michaelov for choosing me to draw for this work in its original context in *The New Yorker*. This job came at a time when I could not have needed it more, and sustained me through a difficult season of my life. Thanks are also owed to Molly Birnbaum, Betsy Stromberg, and Elizabeth Kolbert, my collaborators in this work, for asking me to continue what I had begun in the summer that followed. All gratitude to Chad Beckerman and The Cat Agency for their support of my progress.

ACKNOWLEDGMENTS

ABOUT THE CONTRIBUTORS

ELIZABETH KOLBERT is the author of *Field Notes from a Catastrophe: Man, Nature, and Climate Change*; *The Sixth Extinction*, for which she won a Pulitzer Prize, and *Under the White Sky: The Nature of the Future*. For her work at *The New Yorker*, where she's a staff writer, she has received two National Magazine Awards, a National Academies Communication Award, and the Blake-Dodd Prize from the American Academy of Arts and Letters.

Born in Durham, North Carolina, **WESLEY ALLSBROOK** attended the Rhode Island School of Design. Her work has been recognized by The Art Directors Club, The Society of Publication Designers, The Society of Illustrators, American Illustration, Communication Arts, Sundance Film Festival, Venice Film Festival, Raindance Film Festival, the Television Academy, and The Peabody Awards. She writes and draws for print, television, film, games, and immersive media. She's autistic.

FURTHER READING

To write this book, I relied on many other books. These include the following.

Arrhenius: From Ionic Theory to the Greenhouse Effect by Elisabeth Crawford (Canton, MA: Science History Publications, 1996).

Climate Change 2022: Mitigation of Climate Change. Contribution of Working Group III to the Sixth Assessment Report of the Intergovernmental Panel on Climate Change, edited by P. R. Shukla, J. Skea, R. Slade, A. Al Khourdajie, R. van Diemen, D. McCollum, M. Pathak, S. Some, P. Vyas, R. Fradera, M. Belkacemi, A. Hasija, G. Lisboa, S. Luz, and J. Malley (Cambridge, UK, and New York: Cambridge University Press, 2022). doi: 10.1017/9781009157926.

The Great Indian Phone Book: How the Cheap Cell Phone Changes Business, Politics, and Daily Life by Assa Doron (Cambridge, MA: Harvard University Press, 2013).

How the World Really Works: The Science Behind How We Got Here and Where We're Going by Vaclav Smil (New York: Viking, 2022).

The Legacy of Svante Arrhenius: Understanding the Greenhouse Effect, edited by Henning Rohde and Robert Charlson (Stockholm: Royal Swedish Academy, 1998).

Nomad Century: How Climate Migration Will Reshape Our World by Gaia Vince (New York: Flatiron Books, 2022).

Timefulness: How Thinking Like a Geologist Can Help Save the World by Marcia Bjornerud (Princeton, NJ: Princeton University Press, 2018).

The Two-Mile Time Machine: Ice Cores, Abrupt Climate Change, and Our Future by Richard Alley (Princeton, NJ: Princeton University Press, 2000).

For those interested in further exploring some of the topics raised in this book, I recommend the following.

Cadillac Desert: The American West and Its Disappearing Water by Marc Reisner (New York: Viking, 1986).

The Climate Book: The Facts and the Solutions by Greta Thunberg (New York: Penguin Press, 2023).

Electrify: An Optimist's Playbook for Our Clean Energy Century by Saul Griffith (Cambridge, MA: MIT Press, 2021).

Falter: Has the Human Game Begun to Play Itself Out? by Bill McKibben (New York: Henry Holt, 2019).

Global Warming: The Complete Briefing, 5th ed., by John Houghton (Cambridge, UK: Cambridge University Press, 2015).

The Heat Will Kill You First: Life and Death on a Scorched Planet by Jeff Goodell (New York: Little, Brown, 2023).

How to Avoid a Climate Disaster: The Solutions We Have and the Breakthroughs We Need by Bill Gates (New York: Knopf, 2021).

Limits to Growth: The 30-Year Update by Donella Meadows and Jorgen Randers (White River Junction, VT: Chelsea Green, 2004).

Merchants of Doubt: How a Handful of Scientists Obscured the Truth on Issues from Tobacco Smoke to Global Warming by Naomi Oreskes and Erik M. Conway (New York: Bloomsbury, 2010).

The Uninhabitable Earth: Life After Warming by David Wallace-Wells (New York: Crown, 2020).

A Oneworld Book

First published in the United Kingdom, Republic of Ireland and Australia
by Oneworld Publications, 2024

Text copyright © Elizabeth Kolbert, 2024
Illustrations copyright © Wesley Allsbrook, 2024

The moral right of Elizabeth Kolbert and Wesley Allsbrook to be identified as the Author and Illustrator of this work respectively has been asserted by them in accordance with the Copyright, Designs, and Patents Act 1988

All rights reserved
Copyright under Berne Convention
A CIP record for this title is available from the British Library

ISBN 978-0-86154-866-8
eISBN 978-0-86154-867-5

Printed in China

Editor: Molly Birnbaum | Production editor: Ashley Pierce
Editorial assistant: Kausaur Fahimuddin
Art director: Betsy Stromberg | Production designer: Claudia Sanchez
Production manager: Jane Chinn
Copyeditor: Deborah Kops | Proofreaders: Janet Renard and Christina Caruccio

Oneworld Publications
10 Bloomsbury Street
London WC1B 3SR
England

Stay up to date with the latest books,
special offers, and exclusive content from
Oneworld with our newsletter

Sign up on our website
oneworld-publications.com